THE GROWING NEXUS: ENERGY, ENVIRONMENTAL CAUSES AND SOVEREIGNTY

> Environmentalism is a radical belief with a hidden history. It is not in the same position as other ideologies such as socialism, which as Andrew Dobson has perceptively suggested, was founded on liberal claims about equality and liberty, and fed on liberalism's 'apparent failure' to make its own ideals work. Direct action groups are a predictable consequence of a religious ideology. Deep ecology is such an ideology. It forms the radical edge of the Green movement, and the most consistently anti-humanist trend within it.[1]
>
> —Anna Bramwell

The United States enjoys an exemplary record as a long-time global leader in many key strategic attributes. Salient amongst these are unparalleled individual liberties and freedom, universal unalienable human rights, the world's best and most advanced health care, an exemplary free enterprise business foundation, and a proactive approach towards protecting our natural environment. This paper examines whether the growing initiatives to adopt more aggressive and stringent environmental policies, as Ms. Anna Bramwell fervidly advocates in the opening epigraph above, actually threatens the sovereignty of the United States. Two grossly divergent attitudes have formulated over the recent decades regarding our environment, one that takes a more radicalized view that "the sky is falling" and the earth itself is in peril and the polar opposite contention, that "it doesn't really matter" or that there's simply no cause for concern and that the earth and all within it are meant to be used up according to mankind's desire or need. Neither of these extremes is sensible or conducive to our collective pursuit of happiness, economic viability and freedom as a nation. Countless media medium examples exist today that routinely make a stilted and usually one-sided case for our ostensibly fragile environment. Many of these have actually made the argument for the phenomenon known as 'Global Warming' or 'Climate Change'. These theories warn of impending environmental doom for the world unless direct global action is taken to stop or reverse it. On the surface, many of these arguments seem plausible enough and often appeal to our innate sense of wanting to do all we can to protect, preserve and help our environment for the betterment of all and for the future generations who will inherit what we leave them. This monograph will examine the main thesis points further in the sections that follow.

The global warming hypothesis actually originated in 1896 when Svante Arrhenius, a Swedish chemist, developed the theory that carbon dioxide emissions from the burning of fossil fuels would cause global temperatures to rise by trapping excess heat in the earth's atmosphere. Arrhenius understood that the earth's temps are warmed by a process termed the 'greenhouse effect.' While nearly half of the solar radiation reaching the earth's surface is reflected back into space, the remainder is absorbed by land masses and oceans, warming the earth's surface and atmosphere. This warming process radiates energy, most of which passes through the atmosphere and back into space. However, minute concentrations of greenhouse gases such as water vapor and carbon dioxide convert some of this energy to heat and either absorb it or reflect it back to the earth's surface. These heat-trapping gases function similar to a greenhouse: sunlight passes through, however a small amount of radiated heat remains trapped.[2]

This greenhouse effect serves an essential role in preventing the planet from entering a perpetual ice age actually: Remove these greenhouse gases from the atmosphere and the earth's temperature would plummet by about 60 degrees

Fahrenheit (F). However, scientists who have elaborated on Arrhenius's theory of global warming are concerned that increasing concentrations of greenhouse gases in the atmosphere are causing an unprecedented rise in temperatures globally, with potentially ill effect to the environment and human health.[3]

In 1988, the United Nations Environment Program and the World Meteorological Organization established the Intergovernmental Panel on Climate Change (IPCC), comprised of over two thousand scientists responsible for studying global warming's potential impact on climate. According to the IPCC and their synthetic data and computer models, concentrations of carbon dioxide and temperatures are increasing and are directly attributable to anthropogenic causation.[4] Naturally, many of the people who embrace this view believe that the world's governments must seize the urgent initiative to limit greenhouse gas emissions.[5]

Responding to these pressures, recent and growing bodies of skeptical scientists are questioning the validity of this global warming theory. Per these critics, the IPCC bases its predictions for rising global temperatures on faulty computer climate models, which exaggerate the climate's response to carbon dioxide and associated greenhouse gases while failing to accurately reproduce the motions of the atmosphere. Richard L. Lindzen, professor of meteorology at the Massachusetts Institute of Technology (MIT) warns that "Present models have large errors…and are unable to calculate correctly either the present average temperature of the Earth or the temperature ranges from the equator to the poles…Models…amplify the effects of increasing carbon dioxide." Professor Lindzen asserts that if models accurately represented the role of the major greenhouse gas-water vapor in the climate system, that they would predict a warming of no greater than 1.7 degrees Celsius (C) if atmospheric carbon dioxide levels were doubled. This warming is significantly less than the 4 to 5 degrees C temperature increase forecasted by IPCC models under a doubling of atmospheric carbon dioxide.[6]

Global warming agnostics and skeptics alike also argue that natural climate fluctuation, not human activity, is accountable for the past century's rising temperatures. According to S. Fred Singer, a professor of environmental sciences at the University of Virginia, the earth's climate has never been steady and has continually warmed and cooled over the course of geologic time without any assistance attributable to human activities. Singer says, "The human component in recent global warming is thought to be quite small....The climate cooled between 1940 and 1975, just as industrial activity grew rapidly after WWII. It has been difficult to reconcile this cooling with the observed increases in greenhouse gases."[7] Singer also argues that temperature observations since 1979 are in dispute: Surface readings with thermometers show a rise of only about 0.1 degree C per decade, while data from satellites and balloon-borne radiosondes {miniature transmitters} show no warming—with possible indications actually of a slight cooling—in the lower atmosphere between 1979 and 1997. Until the science behind the global warming theory is more mature, Singer and other skeptical scientists advocate enforcing no limitations or caps on the consumption of fossil fuels.[8]

This current landscape in the United States poses a strategic setting that is characterized by a confluence of politicians, big business, the media, scientists, and environmentalists all playing conflicting roles in this global warming debate as public policy collides head-on with special interests and an uncertain, yet complex scientific theory.[9]

In the last decade this seemingly well-intentioned movement has risen in fervor enough that we are seeing a growing divide between the opposite groups and it continues to grow more contentious. These diverse beliefs have resulted in environmental issues becoming one of the primary political issues of our time.

In fact, for some it actually takes on the form of organized protest and overt activism in order to draw public attention and instigate political action and pressure. Activism is clearly 'in vogue' this decade and is growing in popularity, augmented and encouraged by the rapid pace and proliferation of Information Technology. This manifests itself in the popular and ubiquitous social networks that have a global reach and instant leverage towards concern and action for this 'common cause'. Further evidence that activism has grown in popularity is even seen in *Time Magazine.* They recently named "The Protestor" as the 'Person of the Year!' in 2011's annual issue.[10]

Throughout the world, it is generally accepted that our environment is a precious resource and that it should be preserved and protected accordingly. The stark beliefs and forceful quotes as the one at the start of this monograph from Ms. Anna Bramwell however, goes beyond rational and conventional thought and is really quite disturbing. Particularly because of the adverse implications to the concepts of sovereignty and freedom for individuals. She references ecologism, a term she coined in a previous book, *Ecology in the 20th Century, A History.*[11] She defines this essentially, as the doctrine of political ecology. She continues on to say how the deep ecological activists resemble "an angry and disenfranchised child."[12] She explains further about "The metaphors of earth mother and father rapist that permeate deep ecologist literature are Oepidal; they show the urge to kill the father and marry the mother. The aim is to merge with the mother earth, the tactic to use her strength and skills to destroy the interloper."[13] It is certainly reasonable to state that this sort of approach and thinking is anything but cogent. Yet it is illustrative of the flawed logic that unfortunately, many associated with solemn environmentalism ascribe to.

She further explains how deep ecology is an approach to environmental issues first formulated by Norwegian Philosopher, Arne Naess in 1972, but implicit in ecologism from its inception. "It is apocalyptic, anti-political and anti-reform. It has adopted principles of biological equality, and emphasizes the role of humanity as a participant in nature rather than as nature's steward. Hence it is anti-humanism. The deep ecology vision of humanity is as a natural disaster, something like an exterminatory virus."[14]

From this belief, Ms. Bramwell says actions can flow in several ways. For example, the assumption that nature is harmonious and benevolent leads one to believe that it's possible to find solutions for problems and conflicts without harming any species, "Just by giving up unnecessary and extravagant ways of living, and adopting a more sensitive approach. When explored, however, this apparently innocuous picture turns out to mean living at a Third World Level; not the affluent Third World, either."[15]

In other words, these ecologists believe that taking America 'backwards' towards a Third World Standard is not only a necessary course of action, but for all intents and purposes, actually a required course of action in order to be successful. To quote Ms. Bramwell; "It means protectionism, giving up what the deep ecologist believes to be unnecessary fripperies, restricting personal mobility, and introducing the authoritarian structure of a command economy in order to allocate resources."[16]

She postulates the 'types' of lifestyles that are embraced or admired by these ecologists as the American Indians of the 19th Century. "Deep ecologists really would like to introduce these lifestyles. And since the broad masses, in either hemisphere, are unlikely to want to do this without coercion, and population reduction by one means or another is a crucial part of reducing resource consumption to an acceptable level, the problem for deep ecologists is how to use pacific means to make people change, and how far violent means are justified."[17]

It is inarguable to consider these statements as anything less than extreme. Ms. Bramwell and many of those who subscribe to her values and beliefs actually deplore the modern lifestyles most developed nations enjoy, and believe we should pull most developed nations back to the 19th Century in the name of 'saving the environment'. She further stated that this group knows that "the survival of nature is more important than that of humanity."[18] This inane and irrational mantra is most likely why many who do not espouse these beliefs, state that Global Warming is indeed, an agenda. Especially for these groups who started the movement. Furthermore, for some it even takes on the auspices of a religion. Therefore, this anthology will not devote much discussion or debate covering the 'opposite side' to these concepts, as more than 'equal time' and consideration is almost continuously and amply provided throughout the public media forums and venues. The 'other side' is widely known and frequently presented utilizing traditional fallacies in logic that are emotionally based and short on empirical data or evidence. Environmental causes and 'crises' are often presented to illicit visceral and emotional reactions, not predicated upon logic and facts. As the infamous scientific savant, Albert Einstein is known to have said, "If I had only one hour to save the entire World, I would devote 55 minutes to analyzing the problem and just five minutes to saving it!"

As long as a particular cause is seemingly just, the powerful and ubiquitous media boldly assists as an augmenting driving force, increasing the range, scope, breadth, frequency and pace of your messages or causes. So much so, that in affect it becomes a bona fide Information Operations (IO) Campaign. Many in America saw the 'Global Warming Alarmists' and 'Green Movement' as a leviathan propaganda effort, but some experts and politicians, such as U.S. Senator James M. Inhofe (R-Oklahoma.) have come to recognize it as an actual IO Campaign that has become quite compelling and successful.[19] The growing popularity of group think and our public education system have certainly been a factor in its success. The inherent danger associated here is that many of the proclamations made in the name of global warming or climate change have for the most part, gone unquestioned and unchallenged -- yet the environmental activists have clearly and globally called for not only group and individual action, but have also worked hard to ensure legislation is crafted, introduced and passed to support robust and far-reaching restrictions, limitations and inhibitions on Americans everyday freedoms.

Evidence of the IO campaign is ubiquitous; it has permeated into all aspects of the media, print, movies, television and entertainment industry. A clever example is on the cover page of the *'Life'* section of *'The USA Today.'* Here an article fawns the altruistic benefits of actually "reducing your carbon hoof print once a week" by excluding popular meat and even dairy products from your diet on a weekly basis. The article's opening sentence actually proclaims that *"If every American skipped meat and cheese one day a week, environmentally it would be the same as the country driving 91 Billion fewer miles a year.*"[20]

This brief article illustrates how ludicrous and multi-faceted this coordinated environmentalist IO campaign is. It actually had a chart depicting how every pound of animal protein equated to supposed actual miles driven in relations to carbon dioxide emissions. Their feckless source was merely cited as the "Environmental Working Group". Like many such articles, it was short on facts and evidence yet long on guilt and plays upon the fear factor for everyday Americans. This is how the vast environmentalist movement leverages culpability among the population, especially the children, who are especially vulnerable and swayed by this emotional, yet effective propaganda. Kay Johnson Smith of the Animal Agriculture Alliance in Arlington, Virginia aptly stated that there is a "hidden animal-activist agenda" behind some of these groups associated with the report.[21]

Under the auspices of the betterment and protection of the environment, this increasingly influential green movement has prompted the federal government to become far more restrictive and prescriptive in our everyday lives as Americans and consumers - from the types of fuel we put into or use in our vehicles, to the types of household light bulbs we are now allowed to use. As Americans surrender more and more of our everyday freedoms and liberties in the name of environmentalism, it clearly costs us more, inhibits our choices as consumers and producers, denigrates our ability to be as productive and ultimately threatens our sovereignty as a nation. A recent example of this is being broadcast in the news. The legislature passed a law last year that soon goes into effect which requires all consumers to purchase this new 'environmentally friendly' incandescent bulb that will cost approximately $50 each, compared to the normal standard of about one dollar.[22] Moreover, these supposed "new and improved bulbs" contain mercury, a well-known hazard to human health. If one happens to have the misfortune of breaking one inside their home or office, they'll have an immediate environmental hazard to dispose of, not to mention the future complications and hassles that will be forever associated with disposing of these bulbs once they fail, break or reach the end of their service life.

This monograph examines the overarching factors related to the green movement and associated environmental initiatives, U.S. energy policies and the critical nexus to our sovereignty as a Nation. It utilizes several main points that support the thesis, beginning with the growing body of recent evidence refuting the validity of the 'Global Warming Crisis' or 'Global Climate Change' mantra and the associated implications. It also explains the ramifications that the fervent environmentalist movement has brought to our everyday lives as well as long- term impacts to our sovereignty. Additionally, it examines the disturbing trend of the United States Government's exploiting failure in the promotion and pursuit of these environmental causes. Finally, it examines feasible options and recommendations readily available for the United States to pursue in order to ameliorate both the short and long term effects that our recent unfruitful pursuits and initiatives have garnered.

<u>Growing Scientific Community Refuting the Theory of Global Warming</u>

As stated at the opening of this monograph, there is a growing professional scientific community that has become more public and much more vocal about the erroneous and faulty theories and subsequent conclusions many scientists have made regarding global warming. The U.S. energy policy has certainly been attenuated by more restrictive regulations, prohibitions, exclusions and limitations that have all been more pernicious to our energy status and posture than any other single factor. When all of this is postulated as being necessary for the "global good" and as an integral part of a compelling and successful Information Operations campaign, the environmentalists make an emotionally appealing case.

Several sources provide supporting data for this ever-growing and developing topic. In the following section, this paper presents salient evidence that clearly refutes the case for global warming in its entirety. This data is compiled from several and recent media sources.

A very recent editorial in *The Wall Street Journal* featured a consortium of nearly twenty distinguished scientists and engineers who unequivocally contend that there is "no need to panic about global warming." Signatories include Nobel Prize-winning physicist Ivar Giaever, who supported President Obama in the last election but actually resigned publicly from the American Physical Society (APS) with a letter stating "I did not renew my membership because I cannot live with the APS Policy statement: *'The evidence is incontrovertible: Global warming is occurring. If no mitigating actions are taken, significant disruptions in the Earth's physical and ecological systems, social systems, security and human health are likely to occur. We must reduce emissions of greenhouse gases beginning now.'* In the APS it is OK to discuss whether the mass of the proton changes over time and how a multi-universe behaves, but the evidence of global warming is incontrovertible?" [23]

Despite the international IO campaign of the last several decades drumming the message that ever increasing amounts of the "pollutant" carbon dioxide are destroying civilization, a great number of very prominent scientists actually share Dr. Giaever's sage opinion. Moreover, this number of scientific "heretics" is increasing every year. The underlying reason is due to a confluence of inarguable scientific facts.[24]

Foremost among these facts is the absence of global warming for well beyond the last decade now. This pattern is incongruent with the forecasted warming in the twenty-two years since the U.N.'s Intergovernmental Panel on Climate Change (IPCC) commenced publishing projections and indicates the computer models have significantly exaggerated the amount of warming additional CO_2 can cause. "Faced with this embarrassment, those promoting alarm have shifted their drumbeat from warming to weather extremes, to enable anything unusual that happens in our chaotic climate to be ascribed to CO_2."[25]

The fact is that CO_2 is not a pollutant. It is an odorless and colorless gas that actually is exhaled in high concentrations by all human beings, and a critical component of the biosphere's life cycle. Plant life actually grows and thrives far better when more CO_2 is induced, and greenhouse keepers typically increase the concentrations of it by a factor of three or four to obtain better growth. "This is no surprise since plants and animals evolved when CO_2 concentrations were about ten times larger than they are today. Better plant varieties, chemical fertilizers and agricultural management contributed to the great increase in agricultural yields of the past century, but part of the increase almost certainly came from additional CO_2 in the atmosphere."[26]

This editorial further stated that even if we did in fact embrace these inflationary predictions of the IPCC, that drastic greenhouse gas control policies are simply not justified or acceptable economically. In fact, Yale economist William Nordhaus recently completed an investigation considering a plethora of options policy-wise and determined that nearly the greatest benefit-to-cost ratio is attained with a policy that allows fifty more years of economic growth unfettered by greenhouse gas emissions restrictions and control measures.[27]

This article further asserts that though the number of dissenting scientists on record is increasing, that the majority of young scientists secretly state they share professional doubts about the global warming mantra but are too worried to speak out for fear of not being promoted, or even fired.[28] Certainly, this is the antithesis of how science is supposed to work – as by its very nature, it is rooted in hypothesis and theory that are to be constantly monitored, evaluated, tested and ultimately validated or refuted by the scientific community. This core work provides the fundamental basis for countless discoveries in the exciting and evolutionary world of science. Those hard line global warming scientists clearly have mounted a formidable campaign in blinding support of their one international theory. Anyone from their profession who portends a divergent opinion or belief is shunned from the community and even fired or removed from prestigious positions throughout the world of science and academia.[29]

As to the reasons for this great passion over global warming, and why it has become so vexing that the American Physical Society (APS), caused Dr. Giaver's resignation because they refused a rational request from many of its members to remove the word "incontrovertible" from their description on a significant scientific issue. There are surely many, yet the most telling place to start is the old question "cui bono?" Or the contemporary quip, "Follow the money."[30]

> Alarmism over climate is of great benefit to many, providing government funding for academic research and a reason for government

bureaucracies to grow. Alarmism also offers an excuse for government to raise taxes, taxpayer-funded subsidies for businesses that understand how to work the political system, and a lure for big donations to charitable foundations promising to save the planet.[31]

John Coleman, the cable TV *Weather Channel* founder and current television meteorologist in San Diego, California at KUSI, emphatically states in a series of briefs on his climate blog 'ICECAP' "that not only is the theory of global warming or climate change a well orchestrated con, there simply is no scientific consensus about it either". He made many public and written statements that "Global Warming is a Hoax" which gained great attention nationally. KUSI and ICECAP both received a deluge of emails, greater than 90 percent were in his favor, thanking him for his moral courage to speak out on the issue and thanking KUSI for devoting this type of coverage that the networks will not.[32]

Of course, there were also some negative responses, predominantly ad hominem attacks questioning his motivations or agenda, as is typical with this issue. Some actually requested Mr. Coleman to follow-up with even more solid facts in common everyday terms that laymen could readily understand and relate to. He ably stepped up to the challenge and began in earnest to address these factors at his website and blog.[33]

For his foundational premise, he cites the tried and true adage, "if you tell a lie often enough, everyone will believe it." His fear is that this simple tactic of propagating lies is the very foundation he is up against while opposing this Global Warming "frenzy". According to Mr. Coleman, the common theme of the Global Warming foundational lie is their oft repeated claim that "there is consensus among the 2,500 scientists" that comprise the UN's IPCC on Global Warming and "that Global Warming is unequivocal."[34]

Former Vice President Al Gore claims "the debate is over"[35]. Pollsters tell us that approximately 80 percent of all Americans accept the premise of man-made Global Warming as a momentous problem. Mr. Coleman sees this as the seminal challenge. Just how does one refute all the mass media hyperbole and the commonly accepted views of the general public? Coleman's analytical series of briefs endeavor to accomplish that very thing.[36]

From his vast research and experience, he knows categorically that man-made Global Warming is not occurring. He also knows the research backing the Global Warming fear tactics is significantly flawed. This ultimately illustrates the fact that there is no consensus either, despite the United Nation's IPPC (Intergovernmental Panel on Climate Change) meeting in Bali in December, 2007 where much hype and emphasis was conspicuously displayed and Al Gore had just been awarded the Nobel Peace Prize. By all appearances and most reports, this seemed to be a consensus.[37]

Mr. Coleman contends that nothing could be further from the truth. Supporting his position are Mr. John McLean, a climate data analyst based in Melbourne, Australia and Tom Harris, of Ottawa, Canada, the Executive Director of the Natural Resources Stewardship Project. These gentlemen researched the real story of the IPCC and published the results in the Canada Free Press.[38]

They determined that the IPCC is actually divided into three separate working cells. Just one of those cells was tasked to report on the extent and likely causes of historical climate change in addition to projections for out years. Within this cell they established how many scientists actually agreed with their most salient IPCC conclusion; that humans are indeed causing substantial climatic change. According to them, of the purported 2500, only 62 scientists in total reviewed this critical chapter where this statement appears. Moreover, of the 62 expert reviewers for this chapter, just 55 had a vested interest, with just seven expert reviewers who appeared impartial.[39]

This is obviously a far leap from the universally reported "consensus of 2,500 scientists." Another reputable inside source informed Mr. Coleman that although several thousand scientists were consulted for drafting this report, all of them did not concur with its conclusions.[40]

Dr. John W. Zillman, a generally supportive member of the IPPC specifically noted how the IPCC was very scrupulous about insisting that the final decision about accepting certain review comments should reside with chapter lead authors. He quipped that "some Lead Authors ignored valid critical comments or failed to reflect dissenting views. The report was therefore the result of a political rather than a scientific process."[41]

Further evidence that John Coleman cites to refute this notion of consensus regarding Global Warming includes the 1992 Gallup survey of climatologists. Where 81 percent of those surveyed think that global temperatures have not risen during the past 100 years. They were also uncertain as to whether or not, or even why, such warming had occurred. Or, they actually believed that any temperature increases during that period were in the normal range of variation.[42]

Another study in 1997 that 'Citizens for a Sound Economy Foundation' conducted, involved a survey that determined that state and regional climatologists believe that global warming is largely a natural phenomenon by a margin of 44% to 17%. In this same survey, a total of 89% also agreed that evidence exists to suggest that past climate temperature changes have been large and abrupt, without any human influence. A substantial 86% disagreed that current computer modeling technology is significantly sophisticated enough to make conclusive, accurate predictions about future global temperatures. A full 100% agreed that even if there were *no* human beings, that the Earth's climate would still be changing constantly, and 94% of those agreed strongly. [43] Hence, this clearly refutes the anthropogenic theory of Global Warming among many in the scientific community.

Also, a previous president of the National Academy of Sciences compiled a petition that garnered over 19,000 signatories from American scientists. All concur that the science of climate change, and man's role in it, is uncertain. The Petition clearly states "There is no convincing scientific evidence that human release of carbon dioxide, methane, or other greenhouse gasses is causing or will, in the foreseeable future, cause catastrophic heating of the Earth's atmosphere and disruption of the Earth's climate. Moreover, there is substantial scientific evidence that increases in atmospheric carbon dioxide produce many beneficial effects upon the natural plant and animal environments of the Earth."[44] Another independent organization, The European Science and Environmental Forum, published two monographs, where over a few dozen scientists' present studies contradicting the conclusions of the IPCC.[45]

Massachusetts Institute of Technology (MIT) professor Richard S. Lindzen, Ph.D., Alfred P. Sloan Professor of Meteorology, Dept. of Earth, Atmospheric and Planetary Sciences, is one of the prominent scientists among the 11 who constructed the National Academy of Sciences report on global warming in 2001. Dr. Lindzen almost continuously states that there was in fact a vast range of scientific views brought forth in that report, and that the complete report was clear about there not being a consensus, unanimous or otherwise, about long-term climate trends and what causes them.[46]

Those working cells prepping for the IPCC meeting in December 2007 were directed not to even consider any new research papers after those that were accepted by the IPCC in 2005. This resulted in an entire body of later peer-reviewed scientific work that contradicted the claims before the IPCC to never even be considered. Ultimately, this instigated a lengthy list of prominent scientists to craft a protest letter directly to Ban Ki-moon, the Secretary-General of the United Nations on the UN Climate conference in Bali.[47]

This list of over 100 prominent scientists who signed the letter includes many from several different countries, organizations, and distinguished universities from around the globe. Among the most venerable are: Garth W. Paltridge, PhD, atmospheric physicist, Emeritus Professor and former Director of the Institute of Antarctic and Southern Ocean Studies, University of Tasmania, Australia. Also, Ian Plimer, PhD, Professor of Geology, School of Earth and Environmental Sciences, University of Adelaide and Emeritus Professor of Earth Sciences at the University of Melbourne, Australia. William Kininmonth M.Sc., M.Admin., former head of Australia's National Climate Centre and a consultant to the World Meteorological organization's Commission for Climatology. Five more from Australia include, David Evans, PhD, mathematician, carbon accountant, computer and electrical engineer and head of 'Science Speak' and Stewart Franks, PhD, Professor, Hydro climatologist, University of

Newcastle. Don Aitkin, PhD, Professor, social scientist, retired vice-chancellor and president, University of Canberra. R.M. Carter, PhD, Professor, Marine Geophysical Laboratory, James Cook University in Townsville, Australia. Also, Jon Jenkins, PhD, MD, computer modeling – virology, NSW, Australia.[48]

There are several from the Netherlands, including: Hans Erren, Doctorandus, geophysicist and climate specialist, Sittard, and Salomon Kroonenberg, PhD, Professor, Dept. of Geotechnology, Delft University of Technology, as well as Arthur Rorsch, PhD, Emeritus Professor, Molecular Genetics, Leiden University.[49]

University of Pretoria, South Africa Professor Emeritus, William J.R. Alexander, PhD, Dept. of Civil and Biosystems Engineering; Member, UN Scientific and Technical Committee on Natural Disasters, 1994-2000 also signed this letter of protest. As did several notable experts from Canada, including Ian D. Clark, PhD, Professor, isotope hydrogeology and paleoclimatology, Dept. of Earth Sciences, University of Ottawa, and Christopher Essex, PhD, Professor of Applied Mathematics and Associate Director of the Program in Theoretical Physics, University of Western Ontario, as well as David Nowell, M.Sc., Fellow of the Royal Meteorological Society, former chairman of the NATO Meteorological Group, Ottawa.[50]

Even the President of the World Federation of Scientists, A. Zichichi, PhD, Geneva, Switzerland; Emeritus Professor of Advanced Physics, University of Switzerland signed this letter of protest, as did Zbigniew Jaworowski, PhD, physicist from Warsaw, Poland. He is the Chairman of the Scientific Council of Central Laboratory for Radiological Protection.[51]

Additionally, there is now a group of scientists in excess of 400 who have spoken out as Global Warming skeptics in 2007. This is significant enough that it prompted the United States Senate to publish a formal report about it from the Senate Environment and Public Works Committee.[52] This large group of prominent scientists is comprised from more than two dozen countries and they raised substantial objections to major aspects of the so-called "consensus" on anthropogenic global warming. These scientists, many of whom are current and former participants in the UN IPCC, criticized the climate claims made by the UN IPCC and former Vice President Al Gore.[53]

Even some in the establishment media over the past several years appear to be paying more attention to the growing number of skeptical scientists. In the fall of 2007, the Washington Post Staff Writer Juliet Eilperin conceded the obvious, writing that climate skeptics "appear to be expanding rather than shrinking." Many scientists from around the world have dubbed 2007 as the year man-made global warming fears "bite the dust." Furthermore, many scientists who are also progressive environmentalists believe climate fear promotion has "co-opted" the green movement.[54] As this influential report's introduction stated, this "consensus busters" Senate report was well poised to redefine the debate.

Many of the scientists featured in this report consistently stated that numerous colleagues shared their views, but they will not speak publicly for fear of retribution. Atmospheric scientist Dr. Nathan Paldor, Professor of Dynamical Meteorology and Physical Oceanography at the Hebrew University of Jerusalem, author of almost 70 peer-reviewed studies, explained how many of his fellow scientists have been intimidated. "Many of my colleagues with whom I spoke share these views and report on their inability to publish their skepticism in the scientific or public media." Paldor said. Moreover, the report documented how some skeptical scientists have actually faced threats as well as intimidation.[55]

This comprehensive report details how teams of international scientists are dissenting from the UN IPCC's views of climate science. Nations such as Germany, Brazil, the Netherlands, Russia, Argentina, New Zealand and France all have scientists who banded together in 2007 to oppose climate alarmism. Moreover, more than 100 prominent international scientists sent an open letter in December, 2007 to the UN stating that attempts to control climate were "futile."[56]

Of course those who argue profusely for the other side like former Vice President Al Gore feel obligated to formulate personal affronts against these scientists. Gore has claimed that scientists skeptical of climate change are akin to "flat Earth society members" and similar in number to those who "believe the moon landing was actually staged in a movie lot in Arizona."[57]

Dr. John Coleman's extensive website resource, 'ICECAP' is a veritable repository of empirical data, replete with listings of scientists and experts, to include many members of the IPCC whom have posted articles, blogs, papers and comments all refuting the fallacious man-made global warming predictions. Holistically, this all highlights the fallibility of Al Gore's emphatic declaration that "the debate is over" as well as the cacophonous singing among the press corps about the "consensus of scientists". There simply is no scientific consensus and with valid reason – there is no Global Warming.[58]

Adverse Impacts on the U.S. and World Economy.

The ramifications as well as second and third order effects and, even unseen consequences of this powerful movement emerge across the long-established Diplomatic, Informational, Military, and Economic (DIME) spectrum. This section of the paper will examine some of them, but primarily it will address the adverse impacts to our domestic economy and also explore the global economic effects writ large. The majority of the negative impacts appear to stem from all of the substantial efforts that must be undertaken to counter-balance or offset the adverse affects that result from the pursuit of this now questionable theory of global warming.

U.S. Congressman Michael Grimm fully understands many of these negative implications and he regularly advocates that as a nation, we verify global warming with far greater certainty before we destroy jobs.[59] "Global warming is an ongoing topic of debate in the media and academia. I want to be fully informed with accurate information before I accept any proposal that will destroy American jobs or increase the average family's energy bill as many bills pending in Congress will do. I strongly oppose the 'Cap and Trade' bill pending in the U.S. Senate and as Congressman will fight against any such legislation that fails to be financially sound and prudent."[60]

Congressman Grimm signed the 'Contract From America', clause 2. This rejects Cap & Trade: this effectively stops costly new regulations that would increase unemployment, raise consumer prices, and weaken the nation's global competitiveness with virtually no impact on global temperatures.[61] In support of exploring proven energy reserves and thereby keeping energy prices low, he also signed clause 8 in this same contract; "Pass an 'All-of-the-Above' Energy Policy: Authorize the exploration of proven energy reserves to reduce our dependence on foreign energy sources from unstable countries and reduce regulatory barriers to all other reliable domestic as well as foreign sources."[62] All of these actions will serve to ameliorate the continuously rising energy prices Americans are struggling with today.

Congressman Bill Huizenga of Michigan also understands the negative implications associated with the global warming agenda, particularly as it pertains to the pernicious "Cap and Trade" bill. He says; "this bill means putting a cap on American success and trading away our prosperity. I do not believe that further congressional action is needed to address climate change, especially via the job killing "Cap and Trade" legislation. Today's global warming doomsayers simply lack the scientific evidence to support their claims. A host of leaders in the scientific community have recognized that the argument for drastic anthropogenic global warming is no longer based on science, but is being driven by irrational fanaticism. Clear headedness and a moderate temperament are the best antidotes to this kind of rhetoric. As a good steward of our environment, and for national security reasons, I do support the development of alternative, cleaner sources of energy like nuclear, batteries, and solar. However, the market, not the government, needs to drive that transition. President Obama admitted in 2008 that his proposal to cap greenhouse gases "would necessarily cause energy prices to skyrocket." A recent study from the Heritage Foundation confirmed the President's prediction. The Waxman-Markey Emissions Bill would "cost the average American household over $3,000 per year. This legislation is not needed, and the American people cannot afford it."[63]

Another key concept that Congressman Huizenga keenly comprehends is the fact that it is not the federal government's responsibility to create jobs, but more importantly, it is the responsibility of the government to foster an atmosphere that sets the conditions in order for the private sector to be successful and prosper. For America today, the Congressman says, "This means that the federal government needs to loosen the shackles it has placed on businesses by lowering taxes and eliminating the regulatory barriers that are an obstacle to entrepreneurs everywhere."[64]

Another prominent example from recent headlines that directly harms our economy and jobless rate domestically is President Obama's decision not to support the Keystone XL pipeline project with one of America's greatest neighboring allies, Canada.

The President's decision to shutter the Keystone XL pipeline constitutes little sense and lack of foresight. He rejected the proposal supposedly on the basis of insufficient time to assess the environmental impact shortly following a three-year review by the U.S. State Department that found this project would "be the safest in the history of constructed pipelines."[65]

Furthermore, the pipeline actually enjoyed strong bi-partisan backing from Congress as well as the backing from influential groups such as the U.S. Chamber of Commerce and the Laborers International Union of North America. Also, recent polls indicated that over 70 percent of Americans supported the pipeline's completion. This vastly popular acceptance exists because the project would create over 20,000 jobs, provide a reliable source of energy from a long trusted ally, and would alleviate our exposure to price spikes emanating from constrained oil supplies on the global market. Exactly like those ongoing now.[66] The U.S. environment lobbyists fought this project as a proverbial 'line-in-the-sand'. As far as they're concerned, halting this pipeline blocks development of Canada's oil sands, the second greatest source of carbon in the world.

They believe this will foster development of renewable energy and ideally eliminate, or grossly reduce the need for carbon alternatives.[67] These oversimplified beliefs are terribly naïve and are not supported by reality in the modern world. Blocking the project will not reduce the human race's carbon footprint; in fact, it might actually increase now from the affects of carbon shuffling. Even worse, denial of this project alienates one of our key trusted allies and provides an opportunity for one of our nation's most ardent competitors to gain significant regional influence.[68]

Harmful consequences to Americans include a sharp rise for the average price of gas- over twenty cents per gallon within two weeks of project denial. U.S. households spent over $4,000 for fuel in 2011 and current pricing proves that number likely to increase substantially this year.[69] Surely, the Keystone XL pipeline is no panacea against rising oil and gasoline prices, but its completion would have insulated us from the present and future increases.[70]

Rejection of this pipeline directly led to the loss of some 8,500 jobs in construction, a sector that is currently facing an incredible unemployment rate of nearly 18 percent. Per project data, these jobs would have included 2,584 operators, 1,887 Laborers, 1,921 welders, 272 mechanics, and dozens of quality and environmental control supervisors. Additionally, America also lost 8,500 monitoring jobs and 3,000 jobs for the project's thirty pump stations.[71] "All of this is lost over false environmental concerns already disproven by one of the President's most trusted agencies," said David Holt, President of the Consumer Energy Alliance.[72]

These jobs would have epitomized the so deemed "shovel-ready jobs" that the Obama Administration routinely espouses they want to create to quell the jobless rate and aid everyday Americans in this anemic economy. Yet, apparently none of these aforementioned factors topped the need to placate a key constituency that believes America must reject all matter of fossil fuels from its energy portfolio. Due to this decision to placate a few at the expense of many, America's consumers and businesses now must deal with significant uncertainties, and higher costs for everyday goods and services; all because the President failed to stand up for science, facts and reason.[73]

America needs energy from all available sources in order to remain economically competitive and struggling Americans need jobs to keep their heads above water in this currently untenable economic malady we are in. The Keystone XL pipeline would have provided both. Keystone XL's safety and regulatory standards are unparalleled, it has passed rigorous inspections, and it stands to create 20,000 jobs across a plethora of sectors. For all of these reasons and more, its rejection is difficult to comprehend and is something we must work collectively to reverse.[74]

Indiana Governor Mitch Daniels, gave a very measured and cogent rebuttal to President Obama's 'State of the Union Address' this past January. He addressed everything from our national debt and deficit, to social issues and the environment. Just this past month, he aptly stated that he's an agnostic on the science of global warming. "I don't know if the CO_2 zealots are right, but I don't care, because we simply cannot afford to do what they want to do. Unless you want to go broke, in which case the world isn't going to be any greener. Poor nations are never green."[75] Essentially, as he implied during his 'State of the Union' response, many of this Administration's approaches and policies regarding energy and the environment are actually the fast track to insolvency for America during these volatile and austere economic times.

These imperatives for a viable and coherent national energy policy have been at the forefront of American politics for decades now. Another example was seen from former Chrysler Corporation CEO Lee Iacocca, who early in the year 2000 was queried in an interview about what he thought some of the most important issues of that year's presidential election were and said "Well, our lack of a coherent energy policy is a huge issue right now."[76] This is incredibly intriguing, in that the American economy was far more stable and prosperous than it is currently and gasoline was only $1.50 on average at that time, and yet Mr. Iacocca felt compelled to launch into a contentious discussion of presidential politics, with energy policies being top issues for him.[77]

"When I was running Chrysler I tried to get Presidents Carter and Reagan to focus on this, but when foreign oil is cheap it's just too easy to play politics and to ignore the problem of foreign oil dependency" Iacocca continued. "One of these days America is going to end up in big trouble if we don't make some changes…"[78]

It is now twelve years later, and we are indeed in trouble. Has the United States been "playing politics" all of these years? – as yet another spike in foreign oil prices threaten our economy. These substantial price increases have grown so burdensome that the President can't ignore them any longer, or hope they will go away.[79]

But after denying a permit this past January for the expansion of the TransCanada Corporation's Keystone XL Pipeline project, one that would've presumably expanded American oil refining and decreased America's dependency on foreign oil, the President is now stuck defending his obsession with government backed and funded electric cars that don't sell, and failed solar energy companies owned by his campaign donors which have "lost" hundreds of millions of our tax revenues.[80]

This confluence of elements has forced the President to actually begin talking about domestic oil production as though it were actually a good thing. The polar opposite of his usual stance, that stems from his environmentalist and activist constituency. During his 2008 'hope and change' presidential campaign, candidate Obama routinely scoffed at the very idea of producing any more oil domestically. "If everybody in America just inflated their tires to the proper level," he sarcastically quipped in August of that year, "we would actually probably save more oil, than all the oil we'd get from John McCain drillin' right below his feet... or wherever he was gonna drill..."[81]

These ludicrous ideas from the pulpit acerbate the dialog about America's energy challenges and provide no real solutions to the very real energy problems we face. Americans need to demand both an end to the "politics of oil" and a coherent energy policy. Never again should we settle for cynical "tire gauge politics," or "stimulus dollars" for President Obama's solar energy friends and campaign contributors.[82] Congressional leaders should seize the initiative now, while the President lacks credibility on the issue, and begin to acknowledge the obvious; the United States is home to its own substantial oil and gas resources.[83]

There's absolutely nothing wrong or "unfair" about this fact – we've got plentiful resources right here in the U.S. Yet it is currently a violation of federal law to even search for oil in the Pacific Ocean, in the Gulf of Mexico, in the Atlantic Ocean, and in Alaska. Furthermore, it is also against the law to search for oil shale in the continental United States.[84]

These policies are clearly crippling our domestic energy viability and sustainability, particularly in these unstable and uncertain economic times. Reversing or loosening many of these pernicious policies would ameliorate the situation in short order and would most certainly augment our global standing in the world as well. In 2008, when gas prices spiked, President Bush announced a reversal on his father's ban on offshore drilling. Nearly overnight the price of crude oil dropped and gasoline followed suit. It wasn't that there was any more oil being produced, but on an international market, investors knew that more oil would be coming online – not less. The prices dropped.[85]

With the current policies, such as shuttering the Keystone Pipeline, blocking offshore drilling, and minimizing drilling on federal lands, the forecasts all highlight less availability, not more. The prices continue rising. President Obama can give speeches stating that he's going to open up more of America's resources, but the markets and citizens don't believe him, as every policy he sets in place says the opposite. Prices have indeed continued to climb.[86]

Taking immediate affirmative action is the prudent thing for the United States to pursue right now. It is not only feasible, acceptable and suitable, but it is imperative to our economic and energy viability and long-term sustainability. "We must protect the future of our energy from politicians who have interests only in their own agendas and a misinformed public that believes fossil fuels are destroying the world, when they are actually fueling it! We will be dependent upon fossil fuels for a while, and that is fine. We have hundreds and hundreds of years to figure it out. Our fossil fuels should be utilized as long as possible. There simply is no other sensible option."[87]

U.S. Government Exploiting Failure Pursuing Elusive Green Energy Alternatives

As a staunch environmental advocate, President Obama stated in his July 3, 2010 weekly Presidential address, "Today, I'm announcing that the Department of Energy is awarding nearly two billion dollars in conditional commitments to two solar companies. The first company is Solyndra, a leading-edge clean energy company that I visited earlier this year. The second company is Abound Solar Manufacturing, which will manufacture advanced solar panels at two new plants, creating more than 2,000 construction jobs and 1,500 permanent jobs."

As Bill O'Reilly said in his March 2nd 2012 'Talking Points Memo' "Well, those permanent jobs President Obama lauded weren't so permanent. This week, Abound Solar announced it's laying off half its workforce, nearly 200 employees. Another fiscal disaster."[88] "American voters have to wise up and fast! This country is wasting enormous amounts of money. We owe $16 trillion dollars. That figure could rise to $20 trillion over the next few years. But little is being done to stop the madness."[89]

It is also a well-known fact that the first company, Solyndra equally proved to be an abominable green disaster, after huge Presidential endorsements and overtures, as well as substantial financial backing, Solyndra cost the American taxpayers over $500 Million dollars.

Mr. O'Reilly had two guests on his news program the evening when the above quotes were captured, both were Democrat spokesmen, and he allowed them to present the "other side" to his points but they both failed to produce any sound rebuttals to O'Reilly's 'Talking Points Memo.' In fact, Mr. O'Reilly pointedly pressed them both to come up with just one example of a "Green company success story" that resulted from this Administration's initiatives and the trillion dollars spent over the past three-plus years. Neither of them could cite a single example. This is a sad commentary to say the least, but it inculcates the U.S. government's propensity to exploit failure.

An earlier broadcast of the O'Reilly Factor, from February 17, 2012 featured former CNN Anchor, Lou Dobbs as special guest. Mr. Dobbs spoke about the Solyndra debacle and highlighted the problematic situation that U.S. Energy policy was putting America into. Essentially, he articulated how the U.S. is wasting precious time and money in the pursuit of these feckless companies and programs, all in the name of green technology and the perennial quest for the ever elusive 'clean energy.' These pursuits have not led to success and have only reaped much greater debt for the U.S. at a time when we need to be cutting spending across the board, especially wasteful spending on these fiascos like Solyndra.

Mr. Dobbs further pointed out that meanwhile, while these aforementioned failures dig us much deeper into debt, the price of a gallon of gas is up 92 percent since President Obama took office. The U.S. is actually exporting oil, diesel and gas to markets in Asia and China. He stressed the fact that, when the U.S. exports fuel, it ultimately leads to a shortage domestically for the U.S. and ergo, gasoline prices are rising steadily. It is not even a supply and demand issue. Again, this is costing all American consumers extra money. These extra costs now are seen across the spectrum for every item consumers buy in America, as higher gasoline costs must be passed on in the pricing of all goods and services because the entire market place must absorb them.

The same evening of February 17, 2012 on Greta Van Sustren's cable news show, she had Congressman Cliff Stearns (R-Florida) on as a guest along with the former Solyndra CEO, Dean Rogers, who is now the Chairman of the Democratic Party Convention. She had them discussing the further ill effects of yet another colossal failure regarding a 'clean energy company.' This one, 'Pro Logis' was actually a Solyndra linked company that was given another U.S. government $1.4 billion dollar loan, but two months later, it too went bankrupt.

Further egregious evidence of failing, U.S. taxpayer subsidized, renewable energy companies is epitomized with two more companies. Fisker automotive and A123 Systems, an electric car battery company once touted as a economic stimulus "success story" by former Governor Jennifer Granhom, D-Michigan, has laid off 125 employees since receiving $390 million in government subsidies—but is yet still giving out substantial pay raises to the executives in the company.[90]

The company laid off 125 employees and suffered a net loss of $172 million through the first three quarters of 2011, the Mackinac Center for Public Policy reported, observing that the company's primary customer, Fisker Automotive, is also financially struggling. "Yet, this month A123's Compensation Committee approved a $30,000 raise for Chief Financial Officer David Prystash just days after Fisker Automotive announced the U.S. Energy Department had cut off what was left of its $528.7 million loan it had previously received."[91] The DOE gave the battery company $249.1 million in grant money, while the Michigan government provided A123 with another $141 million in tax credits and subsidies, according to the Mackinac Center.[92]

The current Administration's concerted efforts to push the proliferation of electric cars throughout the United States is yet another canard, based upon flawed logic and gratuitous assumptions. A foremost fatal flaw, is the fact that over 50 Percent of

America's electricity comes from coal. Hence, essential the administration is really pushing coal-powered cars – not "clean, renewable energy" as they mislead Americans to believe. Moreover, these electric cars have such a great initial or start-up cost, that consumers must drive significant amounts before the even break-even fiscally, and worse, once they are finished with these cars, the nation ultimately will be faced with miniature nuclear waste dumps to put all of these hazardous electric car batteries into. This administration is constantly shutting down coal plants that provide our electricity and consumers have seen noticeable price increases in their utility bills – exactly as President Obama said they "necessarily would increase…" The residents of New York and California are intimately familiar with "rolling blackouts" as far as electricity is concerned, yet the government believes they should trust in utilizing far greater amounts of electricity just to charge their cars all the time? Electric cars are simply not a sustainable nor viable solution. It is yet another example of "one step up, two steps back" policies for America's energy.

Electric cars are routinely heralded as 'environmentally friendly', but findings from the University of Tennessee, Knoxville, researchers demonstrate that electric cars in China have an overall impact on pollution that could actually be more harmful to health than gasoline vehicles. Chris Cherry, assistant professor in civil and environmental engineering, and graduate student Shuguang Ji, analyzed the emissions and health impacts of five vehicle technologies in 34 major Chinese cities, focusing on dangerous fine particles. What Cherry and his team found defies conventional logic: electric cars cause far greater overall harming particulate matter pollution than gasoline cars. This particulate matter includes acids, metals, organic chemicals, and soil or dust particles.[93]

"An implicit assumption has been that air quality and health impacts are lower for electric vehicles than for conventional vehicles," Cherry said. "Our findings challenge that by comparing what is emitted by vehicle use to what people are actually exposed to. Prior studies have only examined environmental impacts by comparing emission factors or greenhouse gas emissions."[94]

Hence, yet again an example of a paragon of flawed logic and reasoning, despite hype and propaganda, in terms of actual air pollution impacts, electric cars truly are more harmful to public health per kilometer traveled in China than conventional vehicles.[95]

The Heritage Foundation routinely prosecutes studies related to the core functions of our government. In light of all the news pertaining to renewable or 'clean energy' they have written about some of the aspects of U.S. Energy Policy. They have this to say about the U.S. Department of Energy, "Created in the 1970s as one of President Jimmy Carter's bright ideas, the U.S. DOE has seen its mission evolve from basic research and development to spending billions to commercialize technologies that aren't viable yet-and might never be."[96]

Despite funding numerous projects that never achieve success, or even completion, the U.S. Department of Energy has enjoyed a growing budget in excess of $11 billion over the past ten years – a phenomenal 76 percent increase. Since the United States desperately needs to curb, and ultimately, cut spending, the DOE makes a very suitable place to start. Instead of focusing on new sources of energy, the DOE propagates politically correct pet projects at the expense of the American people.[97]

The Way Ahead

Today's volatile and uncertain strategic landscape demands that the United States be properly prepared for any crisis or eventuality it may confront, especially in the near-term. A viable and resilient energy policy is paramount to ensuring America's readiness, both domestically as well as abroad. America's restrictive and constraining energy policies have forced the nation to be more dependent upon foreign sources of oil, fossil fuels, and rare earth minerals and metals that are crucial to technological development and future enhancements. These oppressive energy policies also atrophy our economic strength and credibility domestically as well as on the global stage. America's operational and strategic agility is significantly disadvantaged by these myopic and failure-prone energy policies and practices, especially of the last several years. America's heuristic tendencies always prevailed in the past, particularly in times of crisis or dire need, but our capabilities are reduced now to the point where the nation is vulnerable and America's preeminent stature in the world is at risk. This culminates in the concept of the sovereignty of the United States being put at risk. As defined below, it lends credence to the thesis that America's sovereignty is diminished by incoherent and ill conceived energy policies.

> Sovereignty – 1. Supreme excellence or an example of it; 2. Supreme power esp. over a body politic: DOMINION, SWAY; 3. Freedom from external control: AUTONOMY, INDEPENDENCE {the chief cause of modern war has been the fallacy of absolute *sovereignty* of the national state – J.T. Shotwell}; 4. Controlling influence; 5. One that is sovereign; esp. an autonomous state.[98]

The Promethean business and investment author Peter Bernstein states in the introduction to his book about managing and mitigating risk; "The world's earliest innovators and inventors provided the missing ingredient that successfully propelled science and enterprise into the world of speed, power, instant communication, and sophisticated finance that marks our modern age by aptly utilizing a rational process of risk taking. Their discoveries about the true nature of risk, coupled with the art and science of choice, provide the foundation of our modern market economy that nations around the world are aspiring to join. Even with all its problems and pitfalls, the free economy, with choice as its central theme, has delivered humanity unparalleled access to the best things in life."[99]

Bernstein contends that the ability to clearly define what may occur in the future and to choose among alternatives lies at the very core of contemporary societies. Prudent risk management guides us over a wide range of decision-making, from allocating wealth to safeguarding public health, from prosecuting war to family planning, from paying insurance premiums to wearing a seatbelt, from planting soybeans to marketing cornflakes.[100]

Bernstein asks many insightful questions in this work, such as, "What is it that distinguishes the thousands of years of history from what we think of as modern times? The answer goes well beyond the progress of science, technology, capitalism and democracy."[101] Bernstein went on to explain how ancient history was replete with brilliant scientists, mathematicians, inventors, technologists, and political philosophers. Even hundreds of years prior to the birth of Christ, the skies were mapped, the great library of Alexandria built, and Euclid's geometry taught. "Demand for innovation in technological warfare was as insatiable then as it is today. Coal, oil, iron, and copper have been at the service of human beings for millennia and travel and communication mark the very beginnings of recorded civilization."[102]

Bernstein's book explains how the revolutionary idea which defines the boundary between modern times and the past is the mastery of risk: the notion that the future is more than a whim of the gods and that humans are not just passive before nature. Until men and women found a way across that boundary, the future was actually a mirror of the past. Nobody takes a risk in the expectation that it will fail. Yet with all the debacles the U.S. has reaped in the elusive pursuit of successful 'green or clean energy', it has little to lose by reverting to the tried and true energy policies of the past. When the Soviets tried to administer uncertainty out of existence through government fiat and planning, they choked off social and economic progress.[103]

Without the mastery of probability theory and other tools of risk management, engineers never could have built the great bridges that span our widest rivers, electric power utilities would not exist, homes would still have to be heated by antiquated parlor stoves or fireplaces, children would still be maimed by Polio, no airplanes would fly, and space travel would never have been a reality. In fact, the obituary of Arthur Rudolph – the rocket scientist who developed the Saturn 5 rocket that launched the first Apollo mission to the moon put it this way; "You want a valve that doesn't leak and you try everything possible to develop one. But the real world provides you with a leaky valve. You have to determine how much leaking you can tolerate."[104]

Bernstein incisively paints the tapestry of the correlation between time and risk with the concepts that follow. Time is the dominant factor in gambling. Time and risk are polar opposites, for if there were no tomorrow, there would be no risk, no consequence. Time transforms risk, and the nature of risk is shaped by the time horizon: the future is the playing field.[105]

U.S. Energy Policy must change course in order to secure America's future, time is of the essence. It is imperative for the United States to adopt these same risk management principles to formulate a coherent and viable energy policy. The United States needs to lift the oil drilling moratoriums that President Obama enacted for the Gulf of Mexico, ANWR, Alaska and other key and bountiful oil sources both on and off the shores within our purview. Current U.S. energy policy is so risk adverse that it negatively impacts Americans domestically and attenuates our strategic ability to project power, economically, diplomatically and even militarily.

Endnotes

[1] Anna Bramwell, "The Fading of the Greens: The Decline of Environmental Politics in the West." (Yale University Press, New Haven & London 1994).

[2] http://www.enotes.com/global-warming-article//print (accessed February 17, 2012).

[3] Ibid.

[4] Ibid.

[5] Ibid.

[6] Ibid.

[7] Ibid.

[8] Ibid.

[9] Ibid.

[10] Kurt Andersen "2011 Person Of The Year," *Time*, December 26, 2011, 54.

[11] Anna Bramwell, "Ecology in the 20th Century, A History." (Yale University Press, New Haven & London 1989).

[12] Bramwell, *The Fading of the Greens*, 160.

[13] Ibid.

[14] Ibid., 161.

[15] Ibid.

[16] Ibid.

[17] Ibid.,162.

[18] Ibid.

[19] U.S. Senator James M. Inhofe, "Climate Change Update Senate Floor Statement." http://inhofe.senate.gov/pressreleases/climateupdate.htm accessed February, 18, 2012.

[20] Elizabeth Weise, "Reducing your carbon hoof print once a week," Life Section, *USA Today,* July 18, 2011.

[21] Ibid.

[22] Life Section, Chart Diagram, *USA Today.* March 12, 2012.

[23] Opinion Page, "No Need to Panic About Global Warming," January 27, 2012, *The Wall Street Journal.*

[24] Ibid.

[25] Ibid.

[26] Ibid.

[27] Ibid.

[28] Ibid.

[29] Ibid.

[30] Ibid.

[31] Ibid.

[32] Joe D'Aleo, Icecap, Sunday, November 11, 2007. http://icecap.us/index.php/go/joes-blog/ accessed February 17, 2012.

[33] Ibid.

[34] John Coleman, "There is No Consensus on Global Warming." jcoleman@kusi.com.

[35] Ibid.

[36] Ibid.

[37] Ibid.

[38] Ibid.

³⁹ Ibid.

⁴⁰ Ibid.

⁴¹ Ibid.

⁴² Ibid.

⁴³ Citizens For a Sound Economy Foundation. September-October, 1997. http://www.cse.org/surveyenviroreg100897.htm (accessed 14 March 2012).

⁴⁴ Ibid.

⁴⁵ Ibid.

⁴⁶ Ibid.

⁴⁷ Ibid.

⁴⁸ Ibid.

⁴⁹ Ibid.

⁵⁰ Ibid.

⁵¹ Ibid.

⁵² http://epw.senate.gov/public/index.cfm?FuseAction=Minority.SenateReport (accessed February 25 2012).

⁵³ Ibid.

⁵⁴ Ibid.

⁵⁵ Ibid.

⁵⁶ Ibid.

⁵⁷ Ibid.

⁵⁸ Ibid.

⁵⁹ 2010 House campaign website, grimmforcongress.com, "Issues", Nov 2, 2010.

⁶⁰ "Michael Grimm on Energy & Oil". http://ontheissues.org/NY/Michael_Grimm_Energy_+Oil.htm (accessed February 18, 2012.)

⁶¹ The Contract From America 10-CFA02 on Jul 8, 2010 (accessed February 18 2012.)

⁶² The Contract From America 10-CFA08 on Jul 8, 2010 (accessed February 18 2012.)

[63] "On the Issues" http://huizengaforcongress.com/on-the-issues/ (accessed February 18 2012.)

[64] Ibid.

[65] 'When American Jobs and Opportunity are Political Collateral Everyone Loses'. Townhall.com Holt, David February 26, 2012. (accessed February 26 2012.)

[66] Ibid.

[67] Ibid.

[68] Ibid.

[69] Ibid.

[70] Ibid.

[71] Ibid.

[72] Ibid.

[73] Ibid.

[74] Ibid.

[75] Andrew Ferguson, 'Ride Along with Mitch'. http://www.weeklystandard.com/print/articles/ride-along-mitch? / (accessed February 17 2012.)

[76] Austin Hill, "Tire Gauge Politics: Obama Becomes The Status Quo On Energy." Townhall.com February 26, 2012. (accessed February 26 2012).

[77] Ibid.

[78] Ibid.

[79] Ibid.

[80] Ibid.

[81] Ibid.

[82] Ibid.

[83] Ibid.

[84] Ibid.

[85] Marita Noon, "Fossil Fuels Deserve a Better Credit Rating than US Govt." Townhall.com February 26, 2012. (accessed February 26 2012).

[86] Ibid.

[87] Ibid.

[88] FOXNEWS TV Program Transcript Date: Thu, 03/01/2012 8PM EST. Transcript Show Name: O'Reilly Factor Transcript Talent Byline: Bill O'Reilly

[89] Ibid.

[90] Erika Johnsen, "Yet Another Subsidized Green Venture Lays Off Workers, Awards Bonuses." Townhall.com February 26, 2012. (accessed February 26 2012).

[91] Ibid.

[92] Ibid.

[93] Whitney Heins, "UT Researchers Find China's Pollution Related to E-cars May Be More Harmful than Gasoline Cars," February 13, 2012, link from *The University of Tennessee Knoxville Home Page at* "Headlines," http://www.utk.edu/tntoday/2012/02/13/researchers-find-ecar-emissions-harmful/ (accessed March 5, 2012).

[94] Ibid.

[95] Ibid.

[96] The Heritage Foundation Home Page, http://failedliberalideas.com (accessed February 7, 2012).

[97] Ibid.

[98] Merriam Webster Third International Dictionary, Unabridged. 1986.

[99] Peter L. Bernstein, *Against The gods; The Remarkable Story of Risk* (Publisher Info) September 07, 1996 Introduction Page 2.

[100] Ibid.

[101] Ibid. Page 3.

[102] Ibid.

[103] Ibid. Page 12.

[104] Ibid. Page 2.

[105] Ibid. Page 15.

www.ingramcontent.com/pod-product-compliance
Lightning Source LLC
Chambersburg PA
CBHW081802170526
45167CB00008B/3294